When Things Begin to Work
Why Success Forms When It Does—and Why AI Makes It Visible
Author: Christopher Effgen
Acknowledgment: With assistance from AI language tools
Published in the United States of America

Table of Contents

Introduction

Most explanations of success arrive too late to be useful.

Something works—a company stabilizes, a project takes off, a technology suddenly matters, a career finds its footing—and only then do stories appear. We hear about vision, intelligence, courage, timing. The explanations feel plausible, but they rarely help anyone else repeat the outcome. Even the people inside the success often feel that the story doesn't quite fit what actually happened.

This isn't because success is mysterious.
It's because **the explanation comes after the structure has already formed**.

Before something works, it usually feels unstable, effortful, and oddly resistant to understanding. Decisions don't stick. Progress resets. The same problems return in slightly different forms. People compensate with more work, more care, more analysis. Nothing quite holds.

Then, sometimes, without a dramatic moment, something changes.

The same effort begins to go further. Certain questions stop reopening. Constraints feel clearer. What once required constant attention starts carrying itself. From the inside, it feels like relief. From the outside, it looks like momentum.

Only later does it look inevitable.

This pattern is not specific to business, technology, or any particular field. It shows up wherever something real has ever formed: in scientific disciplines, organizations, relationships, institutions, creative work, and lives. It is the pattern of **stability emerging from repetition**, and it is far more common than the stories we tell about it.

Most people already recognize this pattern, even if they don't have language for it.

They know the difference between pushing and holding.
Between motion and traction.
Between working harder and something actually working.

What they usually don't know is *why* that difference appears when it does—or how to tell, while they are still inside the process, whether something is beginning to stabilize or whether they are simply expending effort inside a system that hasn't learned how to hold yet.

This book is about that distinction.

It is not a guide to success, and it does not offer advice. It does not explain how to make something work, and it does not promise that anything will. Instead, it describes **what has to become stable before success is even possible**, and how that stability can be recognized when it begins to form.

The reason this matters now is not because the world has changed in some unprecedented way, but because one domain—artificial intelligence—has made these dynamics unusually visible.

In AI, the distance between experimentation and consequence is short. Systems scale quickly, fail publicly, and leave detailed traces behind. Decisions about infrastructure, distribution, governance, and constraint shape outcomes in months rather than decades. Success and instability appear side by side, often before anyone has time to invent a convincing story about them.

Because of this, AI companies provide unusually clear examples of how success forms—not because they are special, but because the structure is exposed before it is hidden by narrative.

But this is not a book about AI.

AI is the clearest current lens, not the subject. The same forces described here operate wherever something real begins to work. If you have ever watched an effort finally stop resetting, or wondered why something that "should have worked" never did, you already have the raw experience this book is written for.

What follows is an attempt to make that experience legible.

It traces how success forms when it does, why it is so often explained incorrectly, and how to recognize the difference between movement and structure while you are still inside it. It does this by looking closely at places where the process is visible—especially in AI—and then stepping back to the pattern that applies everywhere else.

Nothing in this book requires belief. Nothing needs to be adopted. There are no steps to follow and no conclusions to defend.

There is only one question that matters:

Has something begun to hold, or is it still being renegotiated every time you touch it?

Once you can see that difference, you will recognize it anywhere.

PART I — WHEN THINGS BEGIN TO WORK

Chapter 1

Why Success Is Always Explained Too Late

Most explanations of success arrive after they are no longer useful.

Something works—a company stabilizes, a project takes hold, a line of research suddenly matters, a life stops feeling precarious—and only then do stories appear. We hear about insight, persistence, courage, timing. The explanations feel plausible. They may even be sincere. But they rarely help anyone else understand what actually happened, and they almost never help anyone repeat it.

This is not because people are dishonest. It is because **the explanation follows the formation of structure**, not the other way around.

Before something works, it usually feels confusing from the inside. Decisions do not stick. Progress resets. The same issues return in different forms. People respond reasonably: they work harder, think more carefully, seek better explanations. They assume the problem is one of understanding or effort.

Often it isn't.

What is missing, in those early stages, is not intelligence or commitment. What is missing is stability. And stability is difficult to recognize while it is still forming.

When success finally appears, it rarely announces itself. There is no clean boundary between "not working" and "working." Instead, there is a gradual shift in how effort behaves. The same actions begin to produce more consistent outcomes. Certain questions stop reopening. Constraints feel clearer, even if they are not yet articulated. From the inside, this feels like relief. From the outside, it looks like momentum.

Only later does it look obvious.

This is why hindsight explanations are so compelling and so misleading at the same time. Once something holds, it becomes easy to point backward and identify the steps that led there. The sequence looks orderly. The causes appear legible. We tell ourselves that if we had only known then what we know now, the outcome would have been predictable.

It wasn't.

The sense of inevitability is produced by the success itself. Once a structure stabilizes, it reshapes how its own history is perceived. Constraints that were once fragile or contested begin to feel natural. Decisions that were once provisional begin to look necessary. What was uncertain becomes invisible.

This creates a persistent confusion: we mistake the clarity that follows success for the clarity that produced it.

People closest to success often feel this mismatch most acutely. They are asked to explain what happened, and they offer stories that sound right but feel incomplete. They remember uncertainty, reversals, false starts, and luck.

They know that what ultimately worked could just as easily not have. Yet the public narrative demands coherence.

So they provide it.

These stories then become templates. Others copy the visible steps, adopt the language, and repeat the rituals. When it doesn't work, the failure is interpreted as personal: not enough effort, not enough vision, not enough discipline. The possibility that the structure was never ready to hold is rarely considered.

This pattern repeats everywhere.

In science, where discoveries are explained as breakthroughs rather than as the convergence of conditions. In organizations, where success is attributed to leadership rather than to stabilized processes.
In personal lives, where periods of ease are framed as virtue and periods of strain as deficiency.

In each case, the explanation arrives after the fact and obscures the conditions that actually mattered.

What gets lost is the most important distinction of all: the difference between **movement** and **working**.

Movement can happen anywhere. It can be sustained by effort, enthusiasm, or pressure. It can look impressive and feel exhausting. But movement does not accumulate unless something underneath it stabilizes.

Working begins when accumulation becomes possible— when effort starts to carry forward rather than dissipate. That shift is subtle, and it is easy to miss if you are focused on outcomes rather than behavior.

This book begins with the claim that success is not mysterious, but it is commonly misunderstood. It forms when certain things stop being renegotiated, when constraints arrive early enough to shape behavior, and when repetition becomes cheaper than reinvention.

Those conditions are rarely visible in advance. They become visible only when we stop telling stories too soon.

The chapters that follow do not attempt to predict success or explain how to achieve it. They describe what has to become stable before success can occur, and how that stability can be recognized when it begins to form.

The reason this matters is not because everyone should succeed, but because misreading success has real costs. It leads people to force what cannot yet hold, to blame themselves for structural misfit, and to imitate outcomes without understanding the conditions that made them possible.

Understanding comes later. Working comes first.

The rest of this book is an attempt to look at that earlier moment—the point at which something is no longer merely moving, but has begun, quietly, to work.

Chapter 2

What It Means for Something to Hold

There is a moment, familiar to most people, when effort changes character.

The work does not stop. Problems still exist. Attention is still required. But something subtle shifts: the same actions no longer feel like they are being erased as soon as they are performed. Progress begins to accumulate. What was once fragile starts to feel dependable.

This is the moment when something begins to **hold**.

Holding does not mean permanence. It does not mean success is guaranteed, or that difficulty has ended. It means that the structure you are working within is no longer collapsing back to zero each time you step away. The work carries itself forward, at least partially. Yesterday's effort still matters today.

Most people recognize this change immediately, even if they can't explain it. They describe it as relief, traction, or stability. They notice that they are no longer bracing in the same way. Certain decisions stop reopening. Certain explanations stop being necessary.

Before this moment, effort feels compensatory. You are constantly filling gaps, smoothing over inconsistencies, reminding others—or yourself—of what was decided. Nothing quite sticks. The work demands vigilance.

After this moment, vigilance relaxes. Not because things are perfect, but because they are no longer brittle.

This difference is not psychological. It is structural.

Something holds when it can be relied upon without constant reinforcement. When it absorbs small shocks instead of amplifying them. When it persists across variation rather than requiring precise conditions to survive.

You can see this in many places.

A relationship begins to hold when trust no longer needs to be proven daily.
A habit begins to hold when it survives a missed day without collapsing.
An organization begins to hold when processes outlast individual attention.
An idea begins to hold when it continues to shape behavior even when no one is actively advocating for it.

In each case, the signal is the same: the thing continues to function when you are not actively holding it together.

This is why holding is often misinterpreted as ease or talent. From the outside, effort appears to have decreased. From the inside, it feels as though the work has finally found a form that fits. The energy required has not vanished; it has stopped leaking.

Before something holds, it is tempting to believe that more effort will eventually produce the same effect. That if you just push long enough, the structure will stabilize. Sometimes this is true. Often it is not.

Effort can compensate for instability, but it cannot replace structure. Compensation is expensive. It requires constant attention, and it fails abruptly when attention is withdrawn.

Structure, by contrast, is what allows effort to become cumulative.

This is why holding is not a matter of intensity. It is a matter of fit.

When fit is present, small actions have disproportionate effects. When fit is absent, even large efforts dissipate. This is not a judgment about worth or capability. It is a property of how systems behave.

Most failures occur not because people stop trying, but because they continue trying in conditions that do not support accumulation. The exhaustion that follows is often misread as personal limitation rather than as a signal that the structure has not yet formed.

Recognizing whether something is beginning to hold is therefore more important than judging how well it is going.

Holding has specific signs.

Effort becomes quieter.
Repetition becomes possible.
The same action produces similar results across time.
The system tolerates small mistakes without unraveling.
Explanation begins to lag behavior rather than lead it.

These signs are easy to overlook because they are not dramatic. They do not announce success. They simply reduce friction.

When something holds, people often say it "feels obvious" in retrospect. This is the same illusion discussed in the previous chapter. Once stability is present, it reshapes perception. What once required explanation now seems self-evident.

But that obviousness is an effect, not a cause.

The purpose of this book is not to teach how to make things hold. It is to describe what holding looks like when it begins to occur, and how to distinguish it from movement that is sustained only by effort.

That distinction matters because it changes how you relate to what you are building. It changes whether you push, wait, adjust, or let something consolidate. It changes whether you interpret strain as growth or as misalignment.

Most importantly, it changes whether you blame yourself for conditions that have not yet stabilized.

Holding is not a guarantee. Structures can decay. Conditions can change. What holds today may not hold tomorrow. But without holding, nothing ever truly works.

The next chapters examine why forcing stability too early often produces fragility, and how success becomes misread when effort is mistaken for accumulation.

Before that, it is enough to notice this: when something finally begins to work, it is rarely because you discovered the right explanation. It is because the structure finally learned how to carry itself.

That is what it means for something to hold.

Chapter 3

Why Force Creates Fragility

When something is not holding, the most natural response is to push harder.

This response is reasonable. It is often rewarded in the short term. Effort produces visible movement. Problems appear to recede. Deadlines are met. The system looks as though it is progressing.

And then, just as often, it resets.

The same issues return. New problems emerge that feel strangely familiar. The work requires the same explanations, the same interventions, the same vigilance as before. What looked like progress turns out to have been temporary compensation.

This pattern is so common that many people assume it is simply the cost of doing anything meaningful. They normalize strain. They learn to admire endurance. They mistake sustained pressure for growth.

What they are usually seeing is not strength, but **fragility**.

Force can move a system that has not yet stabilized, but it does so by substituting continuous input for structure. It fills gaps that have not closed on their own. It masks misfit rather than resolving it.

As long as the force is applied, the system appears to work. When the force is reduced—even briefly—the underlying instability becomes visible again.

This is why forced systems fail abruptly.

They do not gradually decline. They hold until they don't. The moment attention, energy, or resources dip below a critical threshold, everything unravels at once. From the outside, this looks like sudden collapse. From the inside, it feels like exhaustion finally catching up.

The problem is not effort itself. Effort is necessary whenever something new is being formed. The problem is **when effort arrives too early and stays too long**.

Early effort is exploratory. It tests possibilities. It helps discover what fits. Late effort is cumulative. It builds on what already holds.

But effort applied in the absence of fit does not become cumulative. It becomes compensatory.

Compensation is subtle because it often works just well enough to be misleading. A team holds together because a few people absorb the slack. A project advances because someone keeps intervening. A system appears stable because exceptions are handled manually.

From the inside, this can feel heroic. From the outside, it can look like leadership or dedication.

Structurally, it is a warning sign.

Every system has a limited capacity for compensation. When that capacity is exceeded, failure is not gradual. It is discontinuous. The very mechanisms that kept things moving—workarounds, vigilance, informal fixes—become points of failure themselves.

This is why force creates fragility.

By preventing a system from revealing where it does not yet fit, force delays the information needed for structure to form. Misalignment remains hidden. Boundaries remain undefined. Decisions remain provisional.

The cost is paid later, and often all at once.

This dynamic appears everywhere.

In organizations, where culture holds only as long as key individuals intervene.
In research, where progress depends on constant reinterpretation of results.
In personal lives, where stability is sustained through sheer will.
In institutions, where rules proliferate to compensate for unclear purpose.

In each case, the pattern is the same: force keeps things moving, but it also prevents them from learning how to carry themselves.

This is why fragility is often mistaken for ambition.

Systems that rely on force tend to expand quickly, because compensation can scale faster than structure—at least initially. But expansion increases the load on the compensatory mechanisms. The more the system grows, the more effort is required just to keep it functioning.

Eventually, the cost becomes visible.

People burn out. Processes multiply. Exceptions become routine. The system begins to feel brittle, even if it still appears successful from the outside.

At this point, the usual response is to apply more force.

This accelerates the problem.

The alternative is not to withdraw effort entirely, but to **change what effort is doing**.

Effort that allows misfit to surface is different from effort that suppresses it. Effort that tests boundaries and then steps back creates space for structure to form. Effort that continuously overrides boundaries prevents that formation.

The difference is not moral. It is temporal.

Force is appropriate when discovering what might hold. It becomes dangerous when it substitutes for holding.

This is why some systems appear to "suddenly" become resilient when pressure is reduced. It is not because pressure was unnecessary, but because the reduction allowed latent structure to reveal itself. What was previously being held together by effort finally had a chance to settle.

Understanding this distinction changes how you interpret strain.

Not all strain means progress.
Not all ease means stagnation.
And not all collapse means failure.

Often, collapse is simply what becomes visible when force is removed.

The chapters that follow examine why this removal is so difficult, why success feels inevitable only after force is no longer required, and how systems learn—slowly and imperfectly—what they can actually support.

For now, it is enough to notice this: when something requires constant force to function, it has not yet learned how to hold.

And anything that has not learned how to hold remains fragile, no matter how impressive its motion appears.

Chapter 4

Why Success Feels Obvious in Hindsight

Once something works, it becomes difficult to remember that it ever didn't.

Paths that were once uncertain harden into narratives. Constraints that were once contested feel natural. Decisions that were once tentative begin to look inevitable. From the vantage point of stability, the past reorganizes itself.

This is why success feels obvious in hindsight.

The effect is subtle. It does not announce itself as distortion. It feels like clarity. The story simplifies. The noise drops out. What remains is a clean sequence of cause and effect: this decision led to that outcome; this insight unlocked that result.

The problem is not that these stories are false. It is that they are **incomplete in a specific way**.

They describe what became stable as if it had always been so.

Before success forms, the same elements exist in a much looser configuration. Possibilities overlap. Outcomes are underdetermined. Small differences in timing or context matter enormously. Effort is exploratory rather than cumulative. The system is sensitive to perturbation.

After success forms, those same elements are held together by structure. Sensitivity drops. Variation is absorbed. The system tolerates mistakes that would previously have been fatal.

From the inside, this shift feels like relief. From the outside, it looks like inevitability.

This change in perception is not a failure of memory. It is a consequence of how stability reshapes interpretation.

Once a structure holds, it exerts backward pressure on how its own history is understood. The mind fills in coherence where there was once contingency. What was fragile is remembered as deliberate. What was improvised is remembered as planned.

This happens everywhere.

Scientific discoveries are remembered as breakthroughs rather than as the convergence of methods, tools, and prior failures.
Organizations are remembered as visionary rather than as collections of provisional decisions that finally stopped resetting.
Personal turning points are remembered as choices rather than as moments when circumstances aligned just enough to make a different path viable.

In each case, the sense of obviousness is an artifact of success, not its cause.

This is why attempts to copy success so often fail. The visible steps are real, but the invisible conditions that allowed those steps to accumulate are no longer apparent. What looks like a clear path in retrospect was navigated through uncertainty, compensation, and partial fits that no longer leave traces.

People who attempt to repeat the outcome encounter a different system. They apply the same actions, but the structure that made those actions effective is absent. When it doesn't work, the failure is attributed to execution rather than to missing stability.

The illusion of obviousness thus creates two harms at once.

It misrepresents how success formed, and it misleads those who try to learn from it.

This illusion also obscures the role of timing. Many actions are neither good nor bad in themselves. They are simply early or late. When taken too early, they require force to sustain. When taken too late, they are redundant. Only when conditions are right do they appear prescient.

In hindsight, that timing collapses into judgment. We say an action was smart or foolish, when in fact it was well-timed or poorly timed relative to a structure that was still forming.

Understanding this difference changes how you interpret both success and failure.

It reduces the temptation to moralize outcomes.
It reduces the impulse to force coherence prematurely.
And it makes it possible to recognize when something is becoming viable without needing to justify it immediately.

This is why explanation lags reality.

Explanation requires stability. It requires a form that can be described without dissolving. Before stability, explanation interferes with formation. It freezes elements too early and invites force where patience is needed.

After stability, explanation becomes easy. But by then, it no longer teaches the thing that mattered most: **how to recognize the moment before obviousness sets in**.

The chapters that follow turn toward that earlier moment. They examine the conditions under which renegotiation stops, how stability is staged rather than discovered all at once, and why certain structures begin to repeat while others continue to reset.

For now, it is enough to notice this: when success feels obvious, it is because something has already learned how to hold.

The clarity you feel is not insight leading action.
It is structure reshaping memory.

Understanding that distinction is the first step toward seeing success as it forms, rather than only after it has hardened into story.

PART II — THE GRAMMAR OF STABILITY

Chapter 5

The Three Rooms Every Success Requires

Success never forms in isolation.

This is easy to forget because we often focus on the visible site of work: the person thinking, the team building, the lab experimenting, the company producing. When something finally works, attention collapses onto that focal point. We credit insight, effort, or execution.

What disappears from view are the other places where success had to become possible.

Every success that holds—every structure that repeats—requires at least **three distinct contexts** in which different kinds of work occur. These contexts are not metaphors. They are practical spaces where relevance is staged and decisions bind.

If any one of them is missing or unstable, success remains fragile no matter how strong the work appears.

The first is the place where the work itself happens.

This is where effort is applied. Where ideas are tested, products are built, experiments are run, and relationships are formed. It is the most visible room, and it is the one people usually mean when they talk about "doing the work."

When this room is unstable, progress resets frequently. Results depend on constant attention. Knowledge does not accumulate. The same problems recur because nothing carries forward.

But even when this room functions well, it is not sufficient on its own.

The second is the place where the work reaches beyond itself.

This is where outcomes are taken up by others—where a product finds users, where an idea spreads, where a decision becomes operational, where trust extends beyond the immediate participants. Without this room, success remains local and provisional.

Many efforts fail not because the work is flawed, but because this second room never stabilizes. Distribution is inconsistent. Adoption is fragile. Every extension requires explanation, persuasion, or special handling.

When this room begins to hold, something changes. Feedback becomes continuous. Use becomes habitual. The work starts to matter even when its creators are not present.

The third is the place where constraints arrive early enough to shape behavior.

This room is often the least visible and the most neglected. It is where limits are established before failure forces them into existence. Where boundaries are clarified while choices are still open.

Constraints are commonly experienced as obstacles, but without them success cannot stabilize. A system without early constraints is forced to improvise boundaries later, under pressure. Those late constraints arrive as crises rather than as structure.

When this third room is absent, systems oscillate. They advance and retreat. They expand and then contract. They appear to make progress and then collapse back into renegotiation.

When it is present, constraints feel oddly enabling. Decisions become clearer. Trade-offs become explicit. The system learns what it can and cannot support.

These three rooms serve different functions, but they must align.

The first produces work.
The second allows that work to persist beyond its point of origin.
The third shapes the work before it hardens into something brittle.

Success forms when all three begin to stabilize at the same time.

This alignment rarely happens all at once. More often, one room stabilizes first and pulls the others along. Sometimes work advances faster than distribution. Sometimes constraints arrive before capacity. Sometimes distribution exists but the work itself remains unstable.

In each case, the imbalance creates strain.

Work without distribution feels futile.
Distribution without work feels hollow.

Work and distribution without early constraints feel chaotic.

Force is often applied to compensate for these imbalances. Extra effort fills gaps. Rules proliferate. Explanations multiply. This can sustain motion for a time, but it does not produce holding.

Holding begins when the rooms stop fighting each other.

When work produces results that distribution can absorb. When distribution provides feedback that shapes work. When constraints arrive early enough to guide both.

At that point, effort becomes cumulative rather than compensatory. Decisions stop reopening. The system tolerates variation without unraveling.

This is why success so often feels like a convergence rather than a breakthrough. People speak of things "coming together." They are describing the moment when these rooms align well enough to allow repetition.

It is also why success is so sensitive to context. The same actions taken in a different arrangement of rooms produce different outcomes. What worked once may fail elsewhere, not because the actions were wrong, but because the rooms were not present or not aligned.

The chapters that follow examine how these rooms stabilize, how repetition compiles into defaults, and why certain domains make this process unusually visible.

For now, it is enough to recognize this: whenever something truly works, it is because work, reach, and constraint have found a way to support one another.

Without that alignment, motion continues.
With it, structure begins to hold.

Chapter 6

Rails: How Repetition Becomes Default

When something works more than once, it begins to leave a trace.

The trace is not always visible. It may not be written down or explicitly acknowledged. But it changes what happens next. The second time requires less explanation than the first. The third time requires less coordination. Gradually, what once demanded attention becomes the path of least resistance.

This is how repetition becomes default.

Defaults are often misunderstood as habits, preferences, or culture. But what matters is not why people choose them. What matters is that they **resolve quickly**. When faced with uncertainty, the system follows the path that has already proven it can carry outcomes forward.

That path is what this book calls a rail.

Rails are not designed. They are compiled.

They form when a sequence of actions produces acceptable results repeatedly under similar conditions. Each successful repetition reduces the need to reconsider alternatives. Over time, the rail becomes the answer the system gives automatically.

This process is efficient. Without rails, nothing accumulates. Every decision would have to be re-derived.

Every action would feel provisional. Systems without rails are trapped in constant reconsideration.

But rails also conceal their own history.

Once a rail is established, it no longer feels like a choice. It feels like "how things are done." The fact that it emerged under specific conditions disappears. The rail begins to define what counts as reasonable, professional, or obvious.

This is why rails are powerful—and why they can mislead.

When conditions remain stable, rails are invaluable. They reduce cognitive load. They allow effort to become cumulative. They make coordination possible at scale.

When conditions change, rails can persist anyway.

This persistence is not stubbornness or ignorance. It is the same mechanism that made the rail useful in the first place. The system follows what resolves fastest, even when the fit has begun to degrade.

This is how drift begins.

Drift does not announce itself as failure. Outcomes still occur. Work still gets done. But the cost of maintaining the rail increases. Exceptions multiply. Workarounds become routine. The system relies more heavily on compensation to keep the rail functioning.

From the inside, this often feels like "things have gotten harder lately." From the outside, it can look like complexity or growth.

In reality, the rail is being asked to carry more than it was formed to support.

This is why force often increases when rails harden. The system doubles down on what has worked before, applying more pressure to make it work again. As long as compensation holds, the rail survives. When compensation fails, collapse appears sudden.

Understanding rails requires holding two truths at once.

First, rails are necessary. Without them, nothing works for long.

Second, rails are historical. They are answers to past conditions, not guarantees of future fit.

The mistake many systems make is treating rails as principles rather than as paths. Principles are meant to be defended. Paths are meant to be re-evaluated when terrain changes.

The difficulty is that rails are rarely questioned while they are still resolving. As long as outcomes appear acceptable, the system has little incentive to examine whether the rail remains appropriate.

This is why early constraint matters so much.

When constraints arrive early—before a rail fully hardens—they shape how repetition compiles. The resulting rails are more likely to be robust. When constraints arrive late, rails form first and boundaries are imposed afterward. That combination produces brittleness.

This pattern is visible everywhere.

In organizations, where informal practices harden before formal policy catches up.
In research, where methods persist long after the questions have changed.
In personal lives, where routines outlive the circumstances that produced them.

In each case, the rail once resolved something real. It is not wrong. It is simply no longer sufficient.

The purpose of noticing rails is not to dismantle them reflexively. It is to understand what they are carrying and under what conditions they formed.

A healthy system does not eliminate rails. It allows for **controlled re-staging**—moments where alternatives can be tested without forcing immediate resolution. It distinguishes between rails that still fit and rails that persist only because nothing else has been allowed to form.

This distinction is subtle, and it cannot be reduced to rules. It requires attention to how effort behaves, how exceptions accumulate, and how much compensation is required to keep repetition intact.

When repetition becomes cheaper over time, a rail is likely well-formed.
When repetition becomes more expensive, a rail may be hardening past fit.

The next chapters turn toward a domain where these dynamics are unusually visible—where rails form quickly, drift becomes legible early, and the cost of misfit is exposed before stories can obscure it.

For now, it is enough to notice this: whenever something "just keeps happening," a rail has formed.

Whether that rail is carrying the system forward—or quietly bending it toward fragility—depends on how it was formed and whether the conditions that shaped it still hold.

PART III — WHY AI MAKES THIS VISIBLE

Chapter 7

Why AI Exposes Structure Faster Than Anything Else

Artificial intelligence is not unique because it is intelligent.

It is unique because it makes structure visible before explanation has time to harden around it.

In most domains, the formation of stability is slow. Institutions take years to settle. Organizations accrete habits gradually. Scientific fields advance through long periods of ambiguity punctuated by occasional consensus. By the time success becomes obvious, its conditions are already obscured by story.

AI collapses that timeline.

In AI, cognition is externalized. Decisions are logged. Failures are public. Systems scale quickly, often before anyone fully understands what they are becoming. Boundaries are explicit because they must be. Drift appears early because nothing is old enough to hide it.

As a result, the grammar of success—what has to stabilize before something can work—appears in compressed form.

This does not make AI special in kind. It makes it special in **tempo**.

In an AI system, work, reach, and constraint collide rapidly. A model is trained. It is deployed. It encounters the world. The consequences return almost immediately, sometimes in the same week. Where human institutions take decades to learn what they can support, AI systems often learn it in months.

This speed strips away comforting illusions.

In AI, it becomes obvious when effort is compensating rather than accumulating. Teams can see when progress resets despite increasing scale. They can observe when a system works only as long as specific people intervene. They can measure how often boundaries arrive late, as refusals or rollbacks, rather than early, as design choices.

Because everything is instrumented, these dynamics leave residue.

Logs show repeated failure modes.
Metrics reveal when iteration gets cheaper—or more expensive.
User behavior exposes whether adoption is stable or fragile.
Infrastructure costs make misfit visible in dollars rather than in sentiment.

In slower domains, these signals are often muted or rationalized away. In AI, they are difficult to ignore.

This is why AI companies often experience a peculiar dissonance. From the outside, they appear fast-moving and successful. From the inside, they feel unstable. Fundamentals are renegotiated frequently. Compute, deployment, governance, and funding remain in flux. People sense that something important has not yet settled, even as output increases.

That sensation is accurate.

AI exposes, rather than conceals, the difference between motion and working.

It also exposes how success actually forms.

When an AI company begins to work, the change is unmistakable. Certain questions stop reopening. Infrastructure decisions persist across cycles. Deployment paths repeat without bespoke handling. Governance constraints arrive early enough to shape behavior rather than interrupt it. The same work produces similar outcomes week after week.

At that point, stories begin to appear.

We hear about vision, alignment, product-market fit, or inevitability. But these explanations are still late. They describe what has already stabilized.

What AI makes visible is the moment before that—when renegotiation slows, when rails begin to form, and when effort starts to carry forward rather than dissipate.

AI also makes visible the cost of getting this wrong.

Because AI systems scale rapidly, misfit scales with them. A fragile structure does not fail quietly. It fails publicly. A boundary that arrives late does not inconvenience a small group. It disrupts an entire deployment. Drift does not remain theoretical. It shows up as degraded performance, rising costs, or loss of trust.

This is why AI has become a focal point for debates about safety, governance, and responsibility. These debates are not new. What is new is how quickly the consequences of unresolved structure appear.

AI is where cognition begins to behave like infrastructure.

Once a system holds, it is built upon. Others rely on it. Their work inherits its assumptions. The system's internal decisions begin to shape external behavior at scale. At that point, the question is no longer whether the system works, but **what kind of stability it has formed**.

This is why AI serves as both a lens and a destination.

As a lens, it reveals the universal process by which structure forms: through repetition, alignment of rooms, early constraint, and the gradual compilation of rails.

As a destination, it forces those structures to persist in the world, shaping cognition beyond the people who built them.

The chapters that follow use AI companies as examples not because they are exemplary, but because they are legible. They show, in compressed time, what success looks like before it becomes story.

What you will see in those examples is not genius or inevitability. You will see stabilization. You will see renegotiation slowing. You will see where boundaries arrived early enough to matter—and where they didn't.

AI does not change how success forms.
It simply makes the process impossible to ignore.

That is why it is the clearest place to look.

Chapter 8

Why AI Companies Work When They Do

AI companies do not begin to work when they become impressive.

They begin to work when **the same problems stop being renegotiated every cycle**.

From the outside, the difference is easy to miss. Output may already be high. Models may already be capable. Funding may already be substantial. But inside the organization, the distinction is unmistakable. Before working begins, everything feels provisional. Afterward, something settles.

What settles is not ambition or intelligence. It is structure.

In the early phases of an AI company, nearly every fundamental is unstable. Compute availability shifts. Deployment paths change. Governance questions reopen. Funding conditions fluctuate. Teams compensate by working harder, coordinating more closely, and intervening manually. Progress occurs, but it is expensive and fragile.

This is not failure. It is exploration.

An AI company begins to work when exploration gives way to repetition—when enough of the environment becomes stable that effort can accumulate rather than reset.

Three things typically have to happen.

First, **access to computation must become repeatable**.

This does not mean unlimited compute. It means that training and inference can proceed without renegotiating access each time. When compute remains uncertain, everything else becomes speculative. Roadmaps lose meaning. Iteration slows. Decisions are deferred because their feasibility is unclear.

Once compute stabilizes—even imperfectly—planning becomes possible. The organization can commit to cycles rather than to one-off demonstrations. Work stops being organized around scarcity and begins to organize around sequence.

Second, **paths to deployment must settle**.

An AI system that cannot reliably reach users remains abstract. Early deployments are often bespoke. Each customer requires special handling. Each release demands justification. Feedback arrives unevenly.

When deployment paths begin to repeat, something changes. The system encounters the world in consistent ways. Usage patterns become legible. Failure modes recur. Feedback shapes development instead of distracting from it.

This is often the moment when people say a company has "found product-market fit." What they are really noticing is that **reach has stopped being improvised**.

Third, **constraints must arrive early enough to shape behavior**.

AI systems operate under multiple forms of constraint: safety expectations, regulatory pressure, reputational risk, contractual obligations, and internal governance. When these constraints arrive late—as interruptions, rollbacks, or crises—they fragment progress. Teams oscillate between speed and caution. Releases feel tentative.

When constraints are staged early—before systems harden—they act as guides rather than brakes. Decisions become clearer. Trade-offs are explicit. The organization learns what it can support and what it cannot.

This is not about being restrictive. It is about being legible.

An AI company works when compute, deployment, and constraint begin to reinforce one another instead of pulling in different directions.

At that point, repetition becomes cheaper.

Training cycles follow recognizable patterns.
Releases become versioned rather than ad hoc.
Governance decisions stop reopening.
Failures teach instead of destabilizing.

Only then does momentum become real.

This is why AI companies often feel unstable long after they appear successful. From the outside, the signs of progress are visible. From the inside, fundamentals are still in flux. People sense that success has not yet learned how to carry itself.

When it finally does, the shift is abrupt.

Meetings change tone. Planning horizons lengthen. Exceptions decrease. The organization begins to behave as though the future is not entirely up for negotiation. People stop asking whether something is possible and start asking how to refine it.

At that point, stories begin to circulate.

We hear about vision, leadership, or inevitability. But these explanations remain downstream of the change that mattered most: **the moment when repetition stopped being fragile**.

This is also why AI companies fail in ways that look sudden. As long as compensation can mask instability, progress continues. When compensation breaks—because compute costs spike, governance intervenes, or deployment trust erodes—the underlying misfit becomes visible all at once.

The collapse feels dramatic. In reality, it was deferred.

Understanding why AI companies work when they do does not make success replicable. It does something more modest and more useful. It allows you to distinguish between movement and structure while you are still inside the process.

AI makes this distinction visible faster than any other domain. It compresses time, exposes boundaries, and leaves residue behind. The examples that follow are not success stories in the usual sense. They are reconstructions of moments when structure began to hold.

They show what had to become stable before explanation was possible.

And they show why, once stability appears, explanation is always too late to guide formation—but just in time to obscure it.

PART IV — SEVEN SUCCESS RECONSTRUCTIONS

Chapter 9

OpenAI: When Distribution Stops Being the Problem

OpenAI did not begin to work when its models became impressive.

It began to work when **getting those models into the world stopped being a recurring negotiation**.

In the early phase, capability advanced faster than structure. Models improved, demonstrations circulated, interest grew—but each step outward required bespoke effort. Compute had to be secured anew. Deployment paths were improvised. Feedback arrived unevenly. Every release carried questions that reached beyond the model itself: Who will host this? Who can use it? Under what conditions? With what guarantees?

Nothing was wrong with this phase. It was exploratory. But it was also expensive.

Progress depended on constant coordination. Success required repeated intervention. What appeared to be momentum from the outside felt fragile from within.

The turning point was not a single release or technical breakthrough. It was the formation of a **durable distribution room**—a place where compute access, deployment, and reach could repeat without being re-argued each cycle.

Once that room stabilized, several things changed at once.

Compute stopped being an existential question. It became a planning variable rather than a gating constraint. Training cycles could be sequenced. Roadmaps could extend beyond the next demonstration.

Deployment stopped being bespoke. Instead of inventing a new path for each use, releases began to move through a known channel. Feedback arrived continuously rather than sporadically. The system encountered the world in consistent ways.

Capital formation aligned with repetition. Funding no longer had to justify each new experiment as a singular event. It could be framed as sustaining a cycle that already worked.

From the outside, this looked like acceleration. From the inside, it felt like relief.

This is what it means for distribution to stop being the problem.

When distribution is unstable, capability remains speculative. Each advance raises new questions about reach, responsibility, and sustainability. When distribution stabilizes, capability can accumulate. The work stops resetting.

At that point, releases become versioned rather than tentative. Governance questions arrive earlier and are handled structurally rather than reactively. The organization begins to behave as though tomorrow will resemble today closely enough to plan for it.

This does not eliminate uncertainty. It relocates it.

OpenAI's success, in this sense, is not the product of a single insight or a particular model. It is the result of **repeated cycles becoming cheap enough to sustain**. Once distribution held, the organization could afford to learn in public without renegotiating its footing each time.

Only after this shift did the familiar stories begin to circulate.

We hear about breakthroughs, vision, inevitability. These explanations are not wrong. They are simply late. They describe a system that has already stabilized.

What is harder to see—and more instructive—is the earlier moment, when the problem stopped being "How good is the model?" and became "How do we refine something that already reaches the world reliably?"

That is the moment when an AI company begins to work.

OpenAI's case shows that capability alone does not produce stability. Nor does attention. Nor does funding in isolation. What matters is whether the path from work to world can repeat without force.

When that path holds, success no longer depends on constant persuasion. It becomes infrastructural.

The next reconstruction examines a different configuration—one in which the problem was not distribution, but translation—and how stability formed there by removing a different kind of friction.

Chapter 10

Google DeepMind: When Research Stops Needing Translation

Google DeepMind did not begin to work when its research became groundbreaking.

It began to work when **research stopped needing to be translated in order to matter**.

For a long time, exceptional results existed alongside friction. Discoveries appeared, papers circulated, benchmarks moved—but each success still had to cross organizational boundaries before it could shape products, infrastructure, or sustained practice. Research and deployment lived in adjacent spaces, connected by effort rather than by structure.

Translation was constant.

Ideas had to be explained. Value had to be justified. Feasibility had to be renegotiated. Each advance carried an implicit question: *How does this become real here?* The answer depended less on the result itself than on who was available to champion it and how much coordination could be sustained.

This phase produced impressive outcomes. It also produced drag.

The change that mattered was not a single algorithm or model. It was the consolidation of research, product, and infrastructure into a **shared room where outcomes could bind without reinterpretation**.

Once that consolidation held, several dynamics shifted.

Research outputs no longer waited for adoption. They entered an environment already prepared to absorb them. Infrastructure decisions could be made in anticipation of known research trajectories rather than in response to isolated breakthroughs. Productization stopped being a separate act of persuasion and became a continuation of the same work.

In practical terms, this meant that results could repeat.

A capability demonstrated once could be developed again under similar conditions. Tooling, compute, and deployment pathways persisted across cycles. The cost of moving from insight to impact dropped—not because the work became easier, but because it stopped being re-explained.

From the outside, this looked like a tightening loop: faster iteration, clearer releases, more visible impact. From the inside, it felt like alignment.

This is what it means for research to stop needing translation.

Translation is necessary when contexts are misaligned. It is expensive when it becomes permanent. When translation disappears, it is not because communication improves. It is because **the work and the place where it lands have learned how to recognize each other**.

Only then can research accumulate.

This shift also changed how constraints behaved. Governance, reliability, and deployment considerations arrived earlier, not as obstacles but as parameters. Research was shaped by what could persist, rather than retrofitted to survive contact with reality.

As a result, success began to feel less episodic. Instead of isolated triumphs, there were families of results—related advances that could be built upon, refined, and extended without reopening foundational questions.

The familiar stories about vision and scale followed later. They described a system that had already learned how to carry research forward without constant mediation.

What is instructive here is not the brilliance of the work, but the removal of friction that had previously consumed effort. When translation ceased to be the bottleneck, repetition became possible. When repetition became possible, stability followed.

Google DeepMind's case shows that success does not require research to slow down. It requires research to land in a context that can inherit it.

The next reconstruction examines a different kind of stabilization—one where the central problem was neither distribution nor translation, but the timing and placement of constraint itself.

Chapter 11

Anthropic: When Boundaries Become Operational

Anthropic did not begin to work when it articulated its concerns about risk.

It began to work when **those concerns became operational**.

In the early stages, boundaries existed primarily as arguments. Questions about safety, misuse, and long-term consequences circulated alongside rapid capability gains. These questions were real and pressing, but they lived in discussion rather than in structure. Each release reopened them. Each advance required renewed justification.

This made repetition difficult.

When boundaries are expressed only as principles, they arrive late—after systems are built, after deployments are planned, after expectations are set. At that point, constraint interrupts progress rather than shaping it. Teams oscillate between speed and restraint. Decisions feel provisional because the conditions under which they are acceptable are never fully settled.

What changed was not the existence of boundaries, but **their placement**.

Anthropic's work began to stabilize when boundaries moved from commentary to gating—when they became part of the process by which releases were decided, rather than reactions to what had already occurred. Constraints were no longer something to argue about after the fact. They became something the system expected to encounter early.

Once that happened, several things became possible.

Capability could advance without constant re-litigation of legitimacy. Releases could be planned around known thresholds rather than negotiated under pressure. Partners could align expectations in advance. The organization could learn from iteration instead of defending each step as an exception.

This is what it means for boundaries to become operational.

Operational boundaries do not eliminate risk. They make risk legible. They convert uncertainty into parameters that can be worked with rather than anxieties that must be managed socially. They reduce surprise by clarifying what kinds of progress will be allowed to persist.

From the outside, this can look like restraint. From the inside, it feels like momentum.

The key distinction is timing. When boundaries arrive early, they shape behavior. When they arrive late, they fragment it. Early boundaries guide exploration. Late boundaries force reversal.

Anthropic's success, in this sense, is not about choosing safety over capability. It is about arranging them so that capability does not have to outrun its own constraints.

Once boundaries held, repetition became cheaper. Decisions stopped reopening. Releases could be versioned.

Learning could accumulate rather than being dissipated by debate.

Only after this stabilization did familiar narratives appear—stories about responsibility, foresight, and values. These narratives describe a system that has already learned how to operate within limits.

What is instructive here is not the content of the boundaries, but their role in the structure. Any system that must negotiate its constraints anew each cycle will struggle to repeat. Any system that stages its constraints early can afford to move.

Anthropic's case shows that success does not come from having the right values. It comes from **embedding values where they do work**—in the places where decisions actually bind.

The next reconstruction looks at a different challenge: how openness, often treated as a risk, can become a source of stability when it is arranged as infrastructure rather than as intention.

Chapter 12

Meta: When Openness Becomes Infrastructure

Meta did not begin to work when it decided to be open.

It began to work when **openness stopped being a gesture and became infrastructure**.

Releasing models, on its own, does not produce stability. It produces dispersion. Capabilities spread outward, interpretations multiply, and responsibility fragments. Without structure, openness amplifies variability faster than it creates inheritance.

For a long time, openness was treated as a risk to be managed or a value to be defended. Each release carried anxiety about misuse, misunderstanding, or loss of control. Boundaries arrived late, as patches or warnings, after the model had already left the building.

What changed was not Meta's willingness to release, but the **way release was organized**.

Openness began to function as a system rather than as an exception. Models were released in families, with clear versions and expectations. Safety constraints were separated from intent and turned into portable components that could travel with the model. Distribution was not left to chance; it was anchored in environments where usage would repeat daily and feedback would be immediate.

Once these pieces aligned, openness stopped being fragile.

This is what it means for openness to become infrastructure.

Infrastructure does not rely on persuasion. It relies on repeatability. When openness is infrastructural, others can build on it without renegotiating its meaning. Defaults form. Patterns propagate. The system begins to carry work forward even when its creators are not present.

Meta's distinctive move was to pair open release with **mass distribution**. Models were not simply published; they were embedded in places where interaction was habitual. That pairing mattered. It allowed feedback to arrive at scale and at speed. It also meant that misfit could not hide. Failures appeared quickly. Boundaries that did not hold were exposed immediately.

This pressure forced clarity.

Safety, governance, and acceptable use could no longer remain abstract. They had to be made legible and portable. Instead of being enforced centrally, constraints were packaged as tools, guides, and auxiliary systems—elements that others could inherit alongside the model itself.

From the outside, this can look like relinquishing control. From the inside, it is a way of **moving control earlier**, into the design of what is released rather than into after-the-fact enforcement.

The result was a different kind of stability.

Models could evolve without each release reopening the same debates. Openness could scale without collapsing into chaos. The organization could learn from use rather than speculate about it.

Only after this infrastructure was in place did the familiar narratives appear—stories about ecosystem leadership, influence, or strategic openness. These stories describe a system that has already stabilized.

What is instructive here is not whether openness is good or bad. It is **what happens when openness is treated as a structural commitment rather than as an ideal**.

When openness is infrastructural, it becomes a source of inheritance. Others can rely on it. Defaults form. Repetition becomes possible. When openness is merely expressive, it remains fragile and contested.

Meta's case shows that success does not come from choosing between control and openness. It comes from arranging openness so that it behaves like a stable surface rather than a constant negotiation.

The next reconstruction turns to a case where stability was achieved through a different route entirely—by removing negotiation through ownership, and by collapsing multiple dependencies into a single, tightly coupled system.

Chapter 13

xAI: When Ownership Removes Negotiation

xAI did not begin to work when its models improved.

It began to work when **negotiation stopped being the default condition of progress**.

In many organizations, advancement depends on alignment across multiple external rooms. Compute must be secured from one place. Distribution must be negotiated with another. Capital arrives conditionally. Each cycle reopens the same questions: Who controls the infrastructure? Where will this run? Who decides what ships? Under what terms?

That constant renegotiation is expensive. It fragments attention. It turns progress into a series of one-off victories rather than a repeatable process.

xAI pursued a different path.

Instead of inheriting rooms built by others, it moved to **own the rooms outright**—compute, distribution, and capital— collapsing them into a single, tightly coupled system. The effect was not subtle. Questions that would normally stall progress were removed from the critical path.

Once ownership was established, several things changed at once.

Compute access stopped being provisional. Training cycles no longer depended on external availability or shifting priorities. Capacity could be planned rather than requested. Iteration could be scheduled rather than opportunistic.

Distribution ceased to be a negotiation. By owning a primary surface where models would be used, deployment became default rather than exceptional. Feedback arrived continuously. Usage patterns stabilized. The system encountered the world without mediation.

Capital formation aligned with infrastructure rather than with milestones. Investment no longer had to justify each step as a discrete event. It sustained an ongoing process whose requirements were already understood.

From the outside, this looked like speed. From the inside, it felt like **quiet**.

This is what it means for ownership to remove negotiation.

Negotiation is necessary when authority is fragmented. It disappears when responsibility is consolidated. That consolidation does not guarantee correctness. It guarantees **continuity**. The system can proceed without repeatedly asking permission to exist.

The cost of this approach is obvious. Ownership concentrates risk. It requires sustained capital. It limits flexibility. Once rooms are built, they must be fed.

But it also produces a particular kind of stability.

When compute, deployment, and feedback live in the same system, learning accelerates. Decisions persist. Rails form quickly because the terrain stops shifting underfoot. The organization can afford to discover what fits by iteration rather than by argument.

This configuration also exposes misfit rapidly. Without external buffers, problems cannot be deferred. If something does not hold, it becomes visible immediately—often publicly. There is little room to hide behind coordination failures or partner constraints.

That exposure is not incidental. It is part of the structure.

xAI's case shows that success does not require consensus. It requires **removing the need for consensus at critical points**. By collapsing dependencies into owned infrastructure, the organization eliminated entire classes of delay and ambiguity.

Only after this consolidation did familiar narratives appear—stories about ambition, scale, or inevitability. These narratives describe a system that has already learned how to proceed without constant renegotiation.

What matters here is not whether ownership is desirable. It is **what ownership does to the flow of work**.

When ownership removes negotiation, repetition becomes possible. When repetition becomes possible, structure forms. And when structure forms, success stops depending on coordination heroics.

The next reconstruction examines a case where stability emerged not from consolidation, but from separation— where allowing two different kinds of success to proceed independently prevented them from undermining one another.

Chapter 14

Mistral AI: When Dual Tracks Stop Competing

Mistral did not begin to work when its models attracted attention.

It began to work when **two different kinds of success stopped competing with each other**.

From the beginning, Mistral faced a structural tension common to many emerging systems. On one side was the pull of openness: releasing models broadly, shaping an ecosystem, allowing others to build freely. On the other side was the need for reliability: enterprise deployment, contractual expectations, and the kinds of constraints that make sustained use possible.

Treated as a single problem, this tension is destabilizing. Openness demands flexibility. Enterprise reliability demands control. Trying to satisfy both through the same pathway forces constant compromise. Each release reopens the question of what the system is for.

Mistral's stabilization came from **separating the tracks rather than reconciling them**.

One track was allowed to remain open. Models could be released as substrates—efficient, capable, and freely available. Their role was to propagate, to be tested in the wild, to shape how others built. Success here meant adoption and reuse, not predictability.

The other track was allowed to harden. A flagship line emerged for enterprise use, with clear versioning, defined expectations, and a stable distribution channel. Success here meant repeatability, not reach.

The crucial move was not technical. It was organizational and structural. Each track was given its own room to stabilize in, with its own criteria for what counted as working.

Once this separation held, several things changed.

Openness no longer threatened reliability, because it was no longer responsible for it. Enterprise deployment no longer constrained experimentation, because it was no longer asked to carry it. Each track could repeat under conditions appropriate to its purpose.

This is what it means for dual tracks to stop competing.

Competition between tracks is not a matter of disagreement. It is a matter of shared load. When two different functions are forced to resolve through the same structure, they undermine each other. When they are allowed to stabilize separately, they can reinforce one another indirectly.

Mistral's open releases shaped the broader environment. They influenced expectations, tooling, and standards. That influence fed back into the enterprise track, where familiarity reduced friction. Meanwhile, the enterprise track provided resources, discipline, and persistence that the open track did not require.

Neither track had to justify the other.

From the outside, this can look like hedging. From the inside, it feels like clarity.

The system no longer had to decide whether it was an open project or an enterprise provider. It was both—but not in the same place, and not at the same time.

Only after this separation did familiar narratives appear— stories about efficiency, positioning, or strategic balance. These stories describe a system that has already learned how to let different forms of success coexist.

What matters here is not the particular mix of openness and control. It is the recognition that **some tensions cannot be resolved by compromise**. They can only be resolved by giving each side room to stabilize without interference.

Mistral's case shows that success sometimes comes not from unifying goals, but from **preventing incompatible goals from sharing a single rail**.

When dual tracks stop competing, repetition becomes possible. And when repetition becomes possible, structure forms—quietly, and without needing to be defended.

The final reconstruction turns to a case where stability emerged by narrowing focus rather than by expanding reach—where success was found by moving constraints to the center instead of treating them as obstacles.

Chapter 15

Cohere: When Enterprise Constraints Move to the Center

Cohere did not begin to work when its models became competitive.

It began to work when **enterprise constraints stopped being treated as obstacles and became the organizing center of the system**.

Many AI efforts approach the enterprise as a delayed destination. They lead with capability, then attempt to retrofit reliability, security, and governance after the fact. Each deployment becomes a special case. Each customer requires reassurance. Progress depends on negotiation rather than inheritance.

Cohere moved in the opposite direction.

From early on, constraints that other systems deferred— data control, privacy, deployment boundaries, lifecycle management—were treated as primary. The question was not how to make models impressive, but how to make them **repeatable inside environments where failure is expensive**.

This shift reorganized everything else.

Models were not released as one-off achievements. They were versioned deliberately, with explicit replacement paths. Deprecation was not treated as a failure or a disruption, but as part of how stability is maintained over time. When something changed, the system made that change legible rather than letting it drift.

Deployment was not framed as universal. It was framed as situational. Models could run where the enterprise needed them to run—inside controlled environments, behind boundaries that already existed. This reduced the cost of trust. Adoption did not require belief in future fixes. It relied on present constraints that were already understood.

Workflows became the focal point.

Instead of asking users to adapt to a model, the system adapted to existing processes. Tools were arranged so that AI behavior could persist without constant supervision. The goal was not novelty, but continuity.

This is what it means for enterprise constraints to move to the center.

When constraints are peripheral, they arrive late and disrupt. When they are central, they shape behavior early. Decisions stop reopening. Expectations stabilize. The system learns what it can support and what it should not attempt.

From the outside, this can look conservative. It lacks spectacle. It does not chase maximal reach or public attention.

From the inside, it feels durable.

Cohere's success did not depend on winning a race for general visibility. It depended on forming a structure that could be relied upon by organizations that value predictability over surprise. Once that structure held, repetition became possible without constant negotiation.

Only after this stabilization did familiar narratives appear—stories about enterprise focus, security, or positioning. These narratives describe a system that has already learned how to carry work forward inside tight boundaries.

What is instructive here is not the choice of enterprise over consumer. It is **the placement of constraint**.

When constraints are treated as core design elements rather than as exceptions to be handled later, systems become easier to reason about. They fail less dramatically. They accumulate trust instead of expending it.

Cohere's case shows that success can come from narrowing rather than expanding—from deciding what will be carried and what will be left aside.

When enterprise constraints move to the center, repetition becomes cheaper. When repetition becomes cheaper, structure forms.

And when structure forms, working begins—quietly, and without needing to be impressive.

PART V — WHAT THIS LETS YOU SEE

(Orientation back to the reader, without advice)

Chapter 16

Why Success Is Hard to Copy

Success is not difficult to observe.

What is difficult to copy is **what made it accumulate**.

When something works, it leaves visible residue: products, practices, partnerships, language, roles. These residues invite imitation. They look like causes because they remain after the fact. But residue is not structure. It is what structure leaves behind.

This is why copying success so often produces movement without traction.

People repeat the steps they can see. They adopt the terminology. They recreate the org chart. They mimic the cadence. They hire the same profiles and chase the same benchmarks. The surface looks familiar. The outcome does not.

The problem is not execution. It is **misplaced similarity**.

Success forms when certain conditions align long enough for repetition to become cheaper than reinvention. Those conditions are local. They depend on timing, context, constraint, and what had already stabilized nearby. When you copy the visible actions without inheriting the conditions that made them work, you are reproducing motion without the substrate that allowed it to accumulate.

This is why success feels transferable and then disappoints.

In hindsight, the path looks clean. Decisions appear deliberate. The order seems natural. But that order was not known in advance. It emerged as renegotiation slowed. The sequence you can see now is the result of stabilization, not its blueprint.

When others attempt to follow it, they encounter a different environment.

Constraints are different.
Expectations have shifted.
Platforms have matured or disappeared.
Boundaries arrive earlier or later.
What once required force may now be redundant.
What once worked effortlessly may now require compensation.

The copy fails not because the idea was wrong, but because the **phase has changed**.

This is why timing matters more than talent.

Actions that were well-timed appear brilliant. The same actions, taken earlier, require force to sustain. Taken later, they are unnecessary or ineffective. In both cases, the action itself is identical. Only the surrounding structure differs.

Because this difference is hard to see, failure is often misattributed. People conclude that they lacked insight, courage, or discipline. Organizations conclude that they mis-executed. In reality, they attempted to force a pattern that had already moved on.

Another reason success is hard to copy is that **conditions rarely travel together**.

In the original case, work, reach, and constraint aligned gradually. Each supported the others. In the copied case, one element is usually missing. Distribution is present but work is not yet stable. Work is strong but constraints arrive late. Constraints are clear but reach is improvised.

The system compensates until it can't.

This is also why copying success often leads to brittleness. When a system adopts a rail formed elsewhere, it inherits answers to problems it has not yet encountered. The rail resolves quickly, but not appropriately. Over time, exceptions multiply. Force increases. Drift sets in.

From the outside, this looks like overreach. From the inside, it feels like working harder to maintain momentum.

What is actually happening is simpler: the system is carrying a solution without carrying the problem it was meant to solve.

Understanding this does not make success replicable. It makes imitation **less misleading**.

It shifts attention away from visible outputs and toward less obvious questions:

- What had already stabilized before this worked?

- Which constraints arrived early enough to guide formation?
- Which negotiations stopped, and why?
- What would have failed if force had been withdrawn sooner?

These questions do not produce instructions. They produce orientation.

This is the most one can reasonably hope for.

Success is not a recipe. It is a moment when structure and timing align. That alignment cannot be manufactured on demand. It can only be recognized, respected, and—when present—allowed to consolidate.

This is why copying success feels so frustrating. You are attempting to recreate a convergence without recreating the conditions that allowed it to form.

The chapters that follow turn back to the reader—not to offer advice, but to describe how the difference between motion and structure can be recognized while it is still unfolding.

Not so success can be copied, but so **force can be avoided where it will only create fragility**.

Chapter 17

How to Tell When Something Is Starting to Hold

The moment when something begins to hold is rarely dramatic.

There is no announcement. No clean milestone. No sudden certainty. What changes is not the appearance of success, but the **behavior of effort**.

Before holding, effort feels fragile. It must be reapplied constantly. Decisions need reinforcement. Progress depends on vigilance. The system advances only as long as someone is actively keeping it together.

When holding begins, effort changes quality.

It becomes quieter.

This does not mean less work is required. It means the same work no longer evaporates when attention shifts. Yesterday's decisions still matter today. Small actions begin to have durable effects. The system starts to remember what it has learned.

This change is subtle enough that it is often missed.

People are trained to look for outcomes—growth, recognition, impact. But holding reveals itself first in **process**, not results. The earliest signals are behavioral, not visible.

One signal is that repetition becomes possible.

The same action, taken under similar conditions, produces similar results. This seems trivial, but it is rare in unstable systems. Before holding, repetition produces variance. Each attempt feels like a new experiment. After holding, repetition confirms.

This is why confidence often follows, rather than precedes, stability. Confidence is a response to repeated confirmation, not a cause of it.

Another signal is that **exceptions decrease**.

In fragile systems, exceptions proliferate. Workarounds become routine. Special cases accumulate. The system still functions, but only because someone is constantly handling edge cases manually.

When holding begins, exceptions still exist, but they stop multiplying. The system absorbs small deviations without unraveling. People spend less time explaining why this case is different and more time refining what already works.

A third signal is that **constraints arrive earlier**.

Before holding, constraints appear as interruptions. They arrive late, as problems to be solved or rules to be imposed after something has already gone wrong. Each constraint feels like friction.

When holding begins, constraints show up earlier and feel oddly clarifying. They shape decisions before commitment hardens. They reduce uncertainty rather than increasing it. People stop being surprised by limits and start planning with them.

This shift is often misinterpreted as loss of flexibility. In reality, it is a gain in legibility.

A fourth signal is that **explanation begins to lag behavior**.

In unstable systems, explanation leads. People justify decisions in advance, articulate intent, and defend direction. Words are used to compensate for the absence of structure.

When holding begins, behavior leads and explanation follows. Things work first. Stories come later. People find themselves describing what has already happened rather than arguing for what should happen.

This reversal is a strong indicator that structure has formed.

There is also a change in how failure feels.

Before holding, failures are existential. They call the entire effort into question. Each setback feels like evidence that nothing is working.

After holding, failures become local. They matter, but they do not threaten the whole. The system can learn without collapsing. This is not because risk has vanished, but because the underlying structure can tolerate correction.

These signals are easy to miss because they do not flatter ambition.

They do not feel like victory. They feel like **less strain**.

This is why many people continue to force systems that are not yet holding. They mistake strain for progress and relief for complacency. They push hardest at the moment when patience would allow structure to reveal itself.

Recognizing holding requires a different orientation.

It means paying attention to what persists rather than to what impresses.
It means noticing whether effort is accumulating or dissipating.
It means observing whether decisions are becoming inherited or are still being re-litigated.

None of this requires belief or commitment. It requires attention.

It is also important to recognize what holding does *not* mean.

Holding does not guarantee permanence. Conditions can change. Structures can decay. What holds now may not hold later. Recognizing holding is not the same as assuming durability.

But without holding, nothing ever truly works.

The difference between forcing and allowing something to hold is not a matter of discipline or restraint. It is a matter of timing. Effort applied at the wrong moment delays stabilization. Effort applied after stabilization accelerates it.

Knowing which moment you are in changes everything.

This is why the earlier chapters emphasized recognition over instruction. You cannot force holding into existence. You can only notice when the conditions that allow it are beginning to align.

When you do, the appropriate response is rarely dramatic. It is often simply to stop interfering—to let repetition continue, to let constraints do their work, and to allow the system to consolidate what it has already learned.

The final chapter turns to the hardest case of all: when nothing is holding yet, and the most accurate action is neither force nor imitation, but patience without passivity.

That distinction, too, has a structure.

Chapter 18

When Nothing Is Ready Yet

There are periods when nothing holds.

Effort is real. Intelligence is present. Possibilities exist. But repetition does not accumulate. Decisions reopen. Progress resets. The same work must be redone each time it is attempted. From the inside, this feels like stagnation or failure. From the outside, it can look like indecision or lack of commitment.

Often, it is neither.

It is simply that **the conditions required for holding have not yet aligned**.

This is one of the hardest situations to interpret accurately, because it resists both action and explanation. The instinct to force is strong. So is the temptation to abandon the effort entirely. Both responses misunderstand what is happening.

When nothing is ready yet, force does not accelerate formation. It substitutes for it. Effort can keep things moving, but it cannot teach the system how to carry itself. The more force is applied, the more fragile the structure becomes. Compensation increases. Exhaustion follows. Collapse is delayed, not avoided.

At the same time, withdrawal can be equally misleading. Stepping away entirely often feels like giving up, as though patience were a failure of nerve rather than a recognition of reality. People fear that if they stop pushing, the opportunity will disappear.

But opportunities that depend on force to exist are not yet opportunities. They are possibilities without support.

This is why waiting is so often misinterpreted. Waiting is associated with passivity, hesitation, or fear. In reality, waiting can be a **structural stance**—a way of allowing conditions to develop without prematurely fixing form.

Waiting, in this sense, is not inactivity. It is attention without interference.

When nothing is ready yet, the work that matters most is not advancement, but **discernment**. Observing which elements persist on their own. Noticing which constraints are emerging. Watching where effort dissipates and where it lingers. Paying attention to small signs of alignment rather than to large promises of progress.

This kind of waiting is uncomfortable because it offers no reassurance. It does not produce milestones. It does not generate stories. It often looks like inactivity to those who expect visible movement.

But it is during these periods that structure begins to form—quietly, without announcement.

Many successes that later appear decisive are preceded by long intervals of apparent stasis. During those intervals, the elements that will eventually align are becoming available: tools mature, infrastructure stabilizes, expectations shift, constraints clarify. None of this feels like progress until it suddenly does.

The mistake is to interpret the absence of holding as a personal deficiency.

When nothing is ready yet, the most accurate conclusion is often that **the system has not learned what it can support**. That learning takes time, and it cannot be rushed without distorting the outcome.

This does not mean all waiting is justified. Some situations persist because misfit is being ignored rather than revealed. The difference lies in whether the system is becoming more legible over time.

If waiting produces clarity—if constraints sharpen, possibilities narrow, and signals become more consistent— then structure may be forming. If waiting produces only repetition of the same confusion, then something else may need to change.

The key is that the change required is not always forward motion. It may be a change in scale, context, or timing. It may involve allowing an attempt to fail early and cleanly rather than propping it up indefinitely. It may involve shifting attention to adjacent conditions rather than to the effort itself.

This is why patience without passivity matters.

Patience, in this sense, is not endurance. It is **refusal to force form before it can hold**. Passivity is disengagement. Patience is restraint informed by observation.

Knowing when nothing is ready yet prevents two common errors.

The first is premature commitment—locking in decisions that will later require force to maintain. The second is premature abandonment—discarding possibilities that have not yet had the chance to stabilize.

Both errors stem from the same misunderstanding: the belief that success is a function of will rather than of alignment.

The final chapter does not resolve this tension. It reframes it.

It asks what happens when we stop treating the absence of holding as a problem to be solved, and instead treat it as information. It turns away from action and toward orientation—toward recognizing the road that exists before deciding whether to walk it.

That recognition does not guarantee success.

But it prevents the most costly mistake of all: mistaking motion for progress when nothing is ready to carry it forward.

PART VI — ENDING

Chapter 19

The Road That Already Exists

There is a temptation, once something becomes clear, to turn that clarity into instruction.

To say what should be done next.
To outline steps.
To draw conclusions that feel actionable.

This book resists that impulse.

Not because action is unimportant, but because action that does not arise from fit rarely holds. Most of the force applied in the world is not wasted because people are careless, but because they are acting in conditions that cannot yet carry what they are trying to build.

What this book has described is not a method for success. It is a way of recognizing **when success becomes possible—** and when it does not.

Some things work.

They work not because they are deserved or inevitable, but because certain conditions align long enough for repetition to become cheaper than reinvention. When that alignment occurs, effort accumulates. Structure forms. Stability appears.

When it does not, force substitutes for fit, and fragility follows.

This is not a moral story. It is a structural one.

The universe does not reward effort in the abstract. It stabilizes what fits. This is true of physical systems, biological systems, cognitive systems, and social systems. It is true of technologies and institutions and lives.

Nothing in these pages promises that alignment will occur where you are. Nothing guarantees that what you care about will stabilize. Nothing suggests that waiting is always the right choice or that force is always wrong.

What this book offers is orientation.

It offers a way to tell whether something is beginning to hold, whether repetition is becoming cheaper, whether constraints are arriving early enough to guide formation rather than interrupt it. It offers a way to distinguish between strain that precedes stabilization and strain that signals misfit.

This distinction matters because misreading it has consequences.

When force is applied where structure has not yet formed, effort dissipates and exhaustion follows. When structure is present but unrecognized, people interfere with what is beginning to work. They explain too soon. They optimize too early. They destabilize what was just starting to hold.

The cost of these errors is not just failure. It is the erosion of trust—trust in judgment, trust in timing, trust in one's own perception of what is happening.

Recognizing the difference between motion and structure restores that trust.

It allows you to stop blaming yourself for conditions that have not yet stabilized. It allows you to stop imitating outcomes without inheriting the conditions that made them possible. It allows you to stop forcing explanations where patience would reveal more.

Most importantly, it allows you to notice the road that already exists.

That road is not marked. It is not announced. It does not promise arrival. It becomes visible only when you stop insisting on creating one.

When something holds, it leaves a path behind it. That path is narrow. It is specific to its conditions. It cannot be widened by will or exported by instruction. But it can be recognized.

Those who recognize it follow it not because they are convinced, but because it carries weight. The work feels different. Effort accumulates. The system remembers what it has learned.

Self-interest is enough.

Nothing in this book asks you to believe in the theory it implies. Nothing asks you to adopt its language or defend its claims. If it has done its work, you will recognize its descriptions without needing to agree with them.

You will see the difference between holding and forcing.
Between repetition and reset.
Between structure and motion.

And when you see that difference, you will see it everywhere.

In projects that finally settle.
In organizations that stop churning.
In ideas that persist without explanation.
In moments when nothing is ready yet, and restraint is accuracy rather than hesitation.

The road does not need to be built.

It already exists, wherever something has learned how to carry itself.

The only question that remains is whether you can see it when it appears—and whether you are willing to follow it without demanding that it explain itself first.

A Note on Use, Not Monetization

This book is not designed to scale through persuasion.

Its value does not increase with performance, repetition, or simplification. It increases when the language is used carefully, in places where structure actually matters and where misreading phase has real cost.

When organizations adopt language to describe stability, constraint, and accumulation—when they train with it, build internal tools from it, or rely on it to decide what *not* to force—the value being created no longer resembles a retail transaction. It resembles infrastructure. And infrastructure is licensed, not sold.

That distinction matters.

Licensing preserves the asymmetry this work depends on. It allows the ideas to operate where they belong—inside systems that already carry consequence—without requiring explanation to be repeated, simplified, or performed. It keeps recognition upstream of instruction.

This work does not want to live on stages, in courses, in certifications, or in motivational formats. Those settings reward fluency over accuracy and repetition over fit. They turn recognition into content, and content into commodity.

If the ideas in this book are useful, they will be useful quietly.

They will show up as decisions that stop reopening, systems that stop compensating, and efforts that finally begin to carry themselves. The appropriate economics follow from that reality: low friction at the surface, real value where structure holds, and restraint everywhere else.

That is not a strategy.
It is simply what fits.